Looking Inside Witch Bottles

By Catherine Pincott-Allen

Looking Inside Witch Bottles

Introductory version co-authored with Dr Alan Massey by Catherine Pincott-Allen

First published in Great Britain 2025

Lost Voices Publishing

ISBN 978-1-7385568-6-1

Typeset using Book Antiqua, Rockwell, Horsbe, Segoe Print and Chiller fonts

www.the-field-detectives.com

Contents

Preface

As a founder member of the Field Detectives, we first met Dr Alan Massey and his wife Dr Sylvia Massey in 2017, as we were about to survey a field referred to us by the landowner who told us that Alan and Sylvia had been investigating the landscape around Goadby Marwood for evidence of prehistoric activity. Alan is a retired chemist from Loughborough University, and as we got to know him, we learned that he is not only an authority on prehistoric flint tools, but also an expert on Witch Bottles. Using his expansive knowledge of organic chemistry, he has scientifically examined and analysed many witch bottles and their contents, the result of which led him to write a book Witch Bottle Analysis.

His publication is, as he says, "To act as an open depository/archive of all my photographs, spectra, SEM images, X-radiographs and elemental analyses over the past twenty-five years of research into "witch bottles".

The purpose of this abridged version of his book is to introduce a broader range of people to witch bottles and Alan's detailed and fascinating world of scientific analysis.

Welcome to Looking Inside Witch Bottles.

Catherine Pincott-Allen

Introduction

What is a witch bottle?

A witch bottle can be described as a glass or stoneware bottle that is believed to trap evil spells cast by witches and reflect them back onto the witch. It was also thought they could protect a person or building from harm. The bottles are usually found buried beneath hearths and floors, in chimneys, fireplaces and windows and are often discovered during renovations of old houses and buildings. They have also been found in churchyards, ditches, and riverbanks and can usually be dated to between the 16th and 18th centuries.

The unusual contents of these bottles can include, urine, pins, iron nails, hair, and fingernail clippings, added by a person who has become ill or fallen upon misfortune and believes that their troubles were caused by a witch's spell. The bottle would be sealed with a cork and then buried, usually upside-down. The personal items would belong to the 'spell victim' and the purpose of the pins and nails would be to irritate the witch as they passed urine to the extent that they would reverse the spell and the 'spell victim' would recover. A medical cure in the days before any type of modern medicine was understood. If you can't explain it, then blame it and who better to blame it on but a 'witch'?

An image of the high vacuum apparatus built by Dr Alan Massey to study reactive compounds which explode with air and moisture. The bench on which it sits is eight feet long.

The Reigate Bottle

In 1993, archaeologist David Williams discovered an onion-shaped wine bottle under the floor of a cottage in Reigate, Surrey. Its shape suggested it was probably made in about 1685, but pottery found close by was of a type only made after 1720. The bottle was already old when it was buried, and the man who made it was probably alive in the 1665 plague and the Great Fire of London. There is a chip on the rim which might have been damaged from use, or since witchcraft is involved here, was the chip deliberately knocked off in an attempt to 'kill' the bottle allowing its magic to act more effectively in the spirit world? It still contained liquid and it was assumed to be wine. It's a good job nobody drank the wine, because it turned out to be urine. The Reigate bottle was sent to Loughborough University to be examined by Dr Alan Massey who said,

> ***"In this case, the curse was intended to make the unfortunate object of it feel as if they were weeing with a bladder full of pins".***
>
> Alan Massey

Blame it on the Witch

The dark clouds of doom descended upon England during the late 1720s. With the collapse of the nation's harvest in 1728 came waves of infections that broke out into outbreaks of the 'suffocating cough', 'catarrh', 'flu', 'whooping cough', 'chicken pox' and 'smallpox'.

Who best to blame all this on but witchcraft and what was the best defence against devil's plagues? Throw the curse back at the witch in a bottle and bury it to keep the home and everyone in it safe.

What was Inside and What the Science Tells us

Urine

How did we know it was urine?

Fresh urine contains hundreds of components, so which ones would prove that the Reigate bottle liquid was urine?

A technique called anion chromatography was used to examine the elements in the liquid. Although most of the organic compounds had decomposed when the bottle was opened, the remaining inorganic elements such as sodium, potassium, calcium, magnesium, chloride, sulphate, phosphate and nitrate were proof enough to say confidently that the liquid had once been a sample of urine.

FACT SHEET

Anion Chromatography

A solution is passed down a column packed with a polymer having positively charged groups attached to its surface; the negative anions in solution stick to the polymer with different tenacities and so exit the column at different times, their passage triggering an electronic signal to a moving chart.

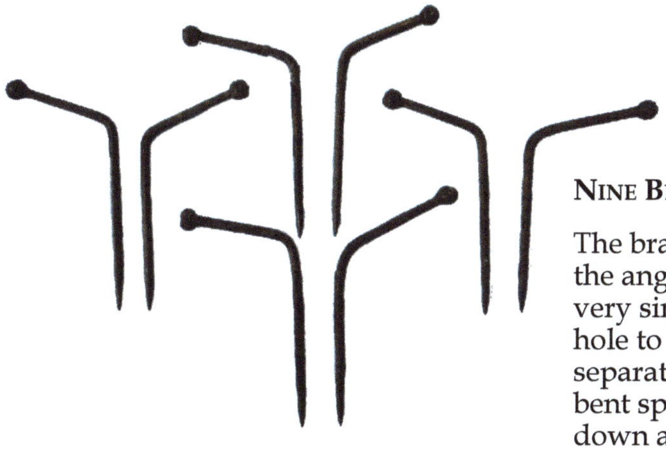

Nine Bent Pins

The brass pins were about 2 cm long, and by measuring the angle of the bends, it was learned that they were all very similar. They would have been placed in a small hole to help in the bending process and each one bent separately. This fact provides evidence that the pins were bent specifically to be added to the witch bottle. Looking down a microscope shows how the head of these handmade pins was constructed by crimping two turns of thin wire around the shaft.

The pins also had patches of what looked like black paint on them, but examination with an electron microscope showed them to be deposits of black copper sulphide. This suggests that a pinch of sulphur, also known as brimstone, had been added to the bottle.

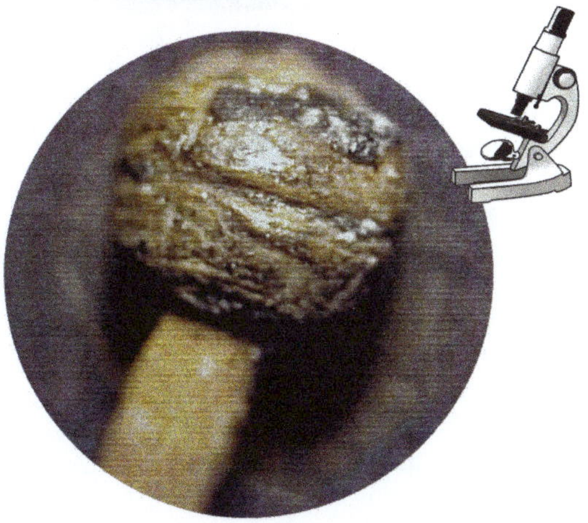

"Remember brimstone spews from the satanic underworld in volcanic fumaroles and, on burning, it produces highly acidic toxic fumes – both admirable reasons for employing brimstone in the counterattack on a witch."

Alan Massey

The Leg Joint of a Spider

The spider may have been added deliberately to scare the witch, but it's more likely it crawled into the bottle accidentally.

The leg came from a common house spider

What Else Was Found?

The Debris

A microscopic examination of the debris in the bottle revealed four pointed hairs, probably eyelashes. One human hair fibre, animal hairs and plant debris.

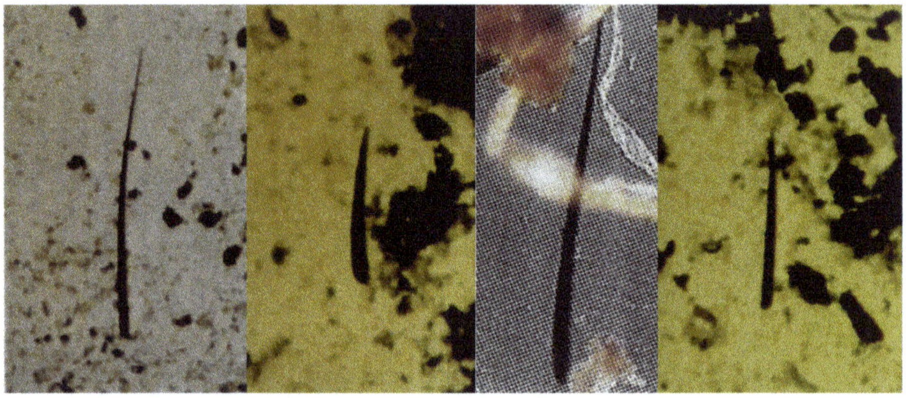

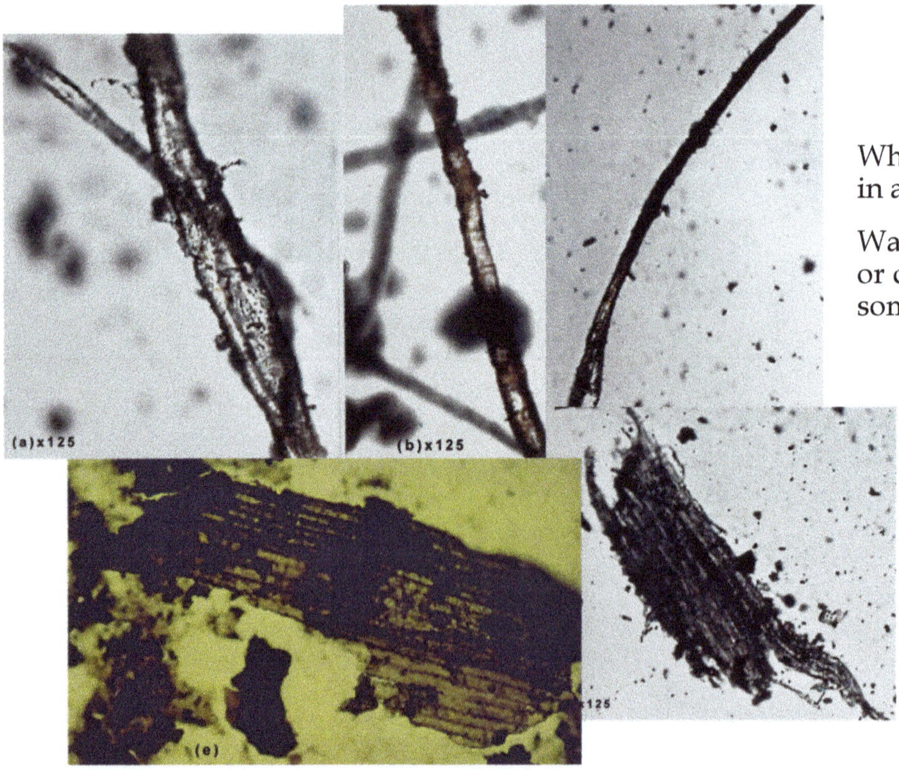

Why should plant debris be found in a witch bottle?

Was the plant added deliberately, or did a seed float into the bottle at some point and start to germinate?

> *"The most amazing component of all witch bottles is the totally invisible one, it is the apparently unswerving belief in the power of magic and witchcraft."*
>
> Alan Massey

Further tests on the solid debris identified calcium carbonate and calcium phosphate. Scanning Electron Microscopy (SEM) was also used and the resulting images revealed that much of the calcium carbonate appeared as tiny spherical clusters of calcite crystals.

Some beautiful single crystals were also visible and other SEM views show a sheaf of needle-like crystals.

FACT SHEET

An SEM uses electrons that are reflected off the near-surface region of a sample to create an image. It can magnify an object 10 times to 3,000,000 times.

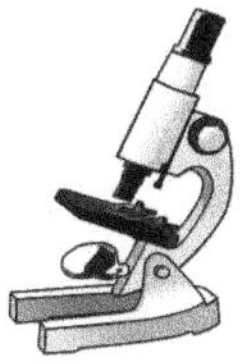

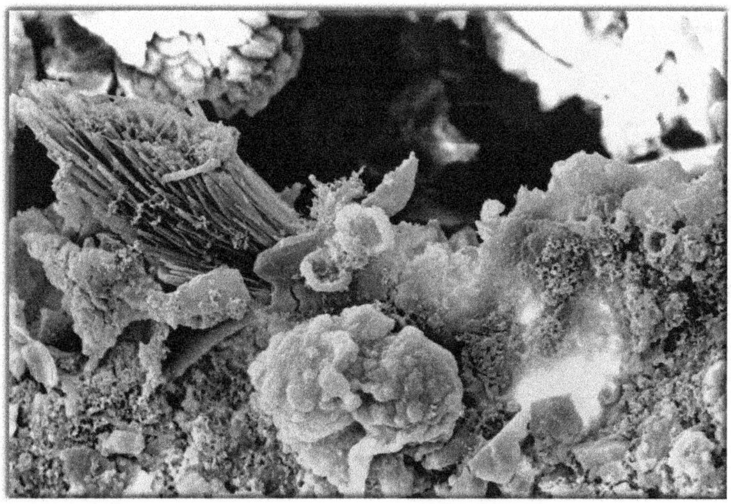

Single crystals of calcium carbonate growing in a microscopic 'cave'.

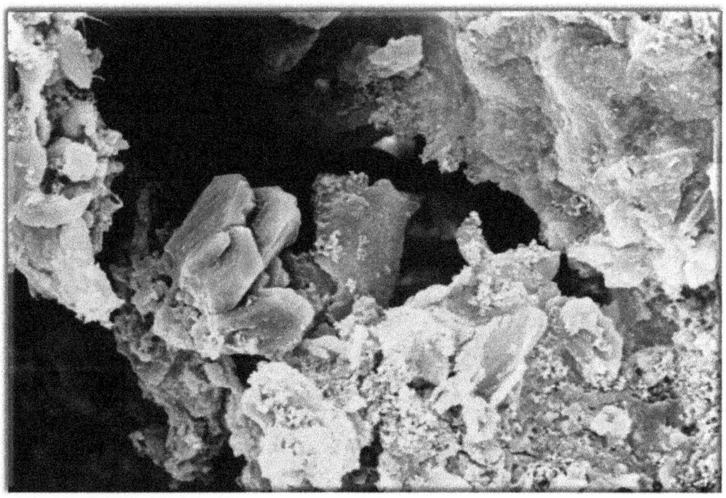

THE FELMERSHAM BOTTLE

Perhaps, shortly after King Charles II died in 1685, a villager from Felmersham, Bedfordshire, secretly buried a sealed bottle upside down in a hole beneath his cottage fireplace. Over three hundred years later, in 2001, it was discovered by a workman during alterations to the cottage. Unfortunately, the cork had been lost a long time ago and any liquid inside had leaked away. Once again, the bottle was examined by Dr Massey and its contents were scientifically studied.

This type of stoneware bottle is known as a bellarmine after Cardinal Bellarmine, but more correctly as a Bartmann or bearded man bottle. It is typical of the salt-glazed stoneware made in the Rhineland for holding wine and this one probably dates from the late 1600s. The bottles came in various sizes, but the Felmersham example is about 23 cm tall.

THE FELMERSHAM BOTTLE TAKING A TURN

What was Inside and What the Science Tells us

Iron Nails & Pins

The bottle was X-rayed and revealed that its neck was completely blocked with a mass of concreted iron nails. The blockage was not particularly thick, and the X-ray image showed at least four pins stuck within the mass. When the blockage was broken with a screwdriver, it split into three lumps and pins protruded out of two of them. The bottle was shaken over some clean paper and one small rusty nail measuring about 14 mm, dropped out. This was all that was left from the ghosts of a few nails that had disintegrated to rust.

Hair

Something soft was still partially blocking the neck… It was a very dirty hank of hair and trapped within it was a bent pin.

There were also two grey hairs from someone else's head. The preparation and burial of witch bottles required a great deal of effort, so perhaps, the bottle was made up by a different person because the victim was too ill to do it themselves and the hairs fell in by accident.

The Washed Hair

Pins

The X-ray images were used to count the number of pins found inside and although this proved quite difficult because they were trapped in the rusty mass, there were probably eleven. Interestingly, four or five of the pins were straight, but the rest were bent and one of these was doubly bent.

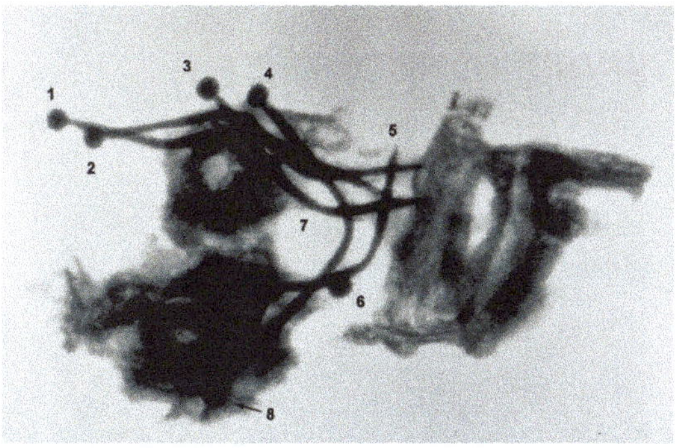

"The amount of rust still attached to the pins contained enough metallic iron to make the pins stick to a magnet".

Alan Massey

Urine

Even though any liquid had dried up, by rinsing the bottle out with distilled water and analysing the wash water by anion chromatography, the presence of chloride, nitrate and sulphate ions proved that urine had once been in the bottle.

Insects

Under the microscope, five tiny springtails barely 1 mm in size were seen. These insects would have crawled into the bottle in recent times or they would have been crushed as the iron nails rusted.

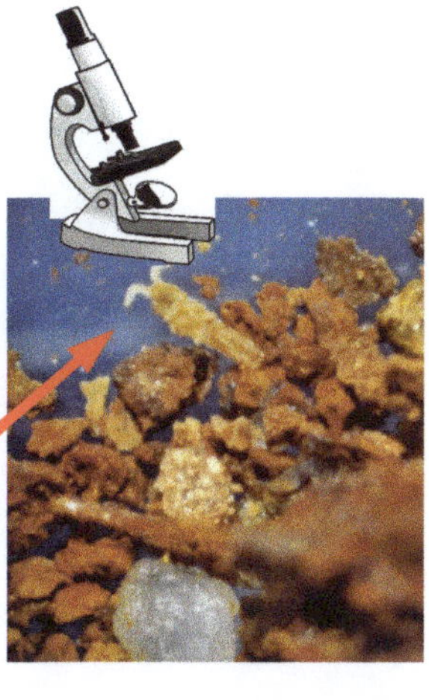

The Greenwich Bottle

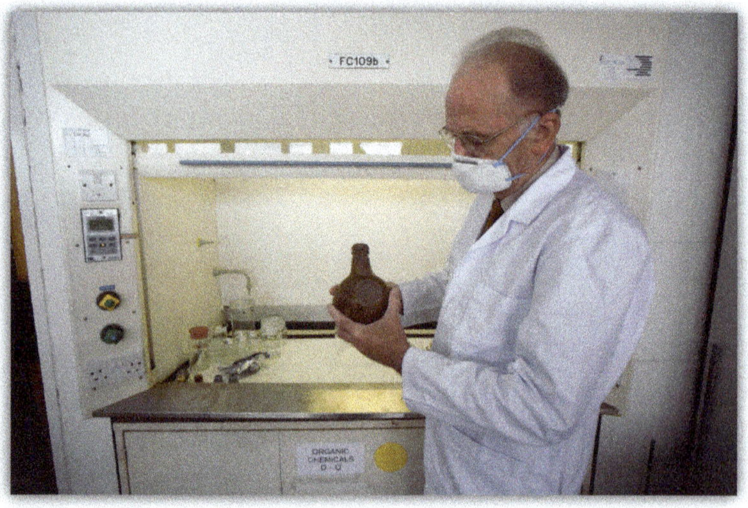

In 2004, a seventeenth-century bellarmine was discovered in Greenwich, London, during building work on the east side of Church Street. What was different about this particular bottle was that the cork was still intact. When it was swirled gently, it was obvious that a liquid was inside and some loose, solid material. The intrigued owners contacted Loughborough University and Dr Massey began the first, full scientific analysis of a complete witch bottle.

The bottle is also about 23 cm tall and to find out what was inside, it was X-rayed and also passed through a CT scanner.

X-RAY

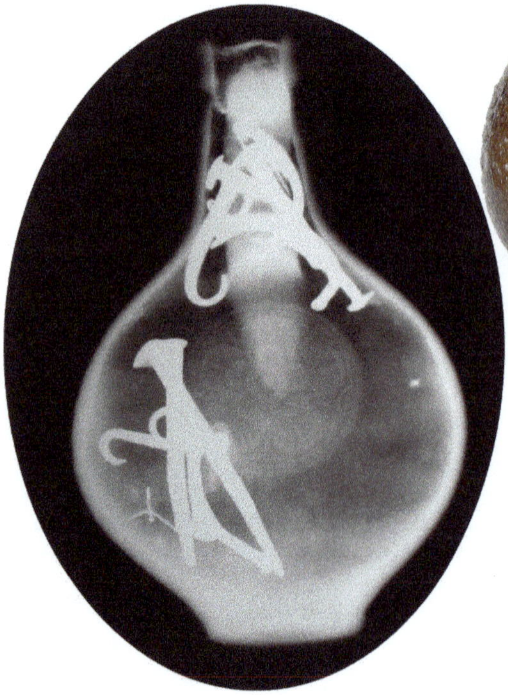

CT SCAN

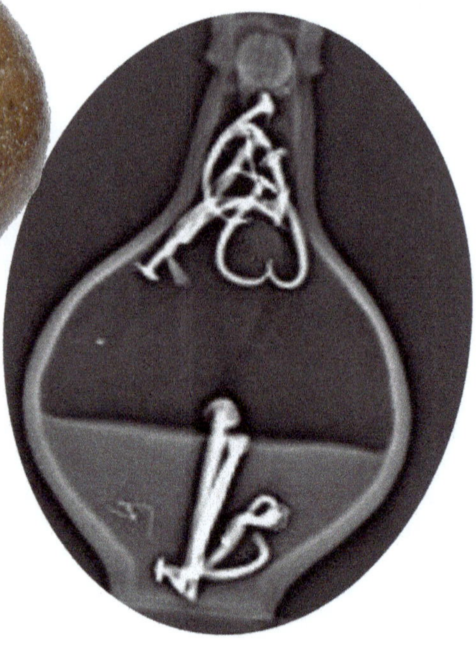

What was Inside and What the Science Tells us

The X-ray and CT images clearly showed many large nails and pins trapped in the neck of the bottle, indicating that the bottle had been deposited upside down. It could also be seen that both the nails and pins were bent, and the CT scan revealed the presence of a liquid. We were clearly dealing with a bellarmine that had been prepared as a witch bottle. Maybe the fearsome face on the bottle's neck also acted as a counterattack on the witch.

"The beast and the false prophet were cast alive into a lake burning with brimstone."

Revelations 19:20

Brimstone

The bottle was opened under strict scientific conditions and several small samples of the liquid were taken for analysis. A black solid was seen floating in the liquid and a simple scientific test proved it to be iron sulphide formed by a chemical reaction between the iron nails and sulphur. Once again, suggesting that brimstone had been added to the bottle.

Urine

Analysis of the organic acids in the liquid by GC-MS (Gas Chromatography-Mass Spectrometry) confirmed it had been human urine. One of the compounds detected in the urine was cotinine - a metabolite of nicotine. This strongly suggests that the person who produced the urine sample was a smoker.

Fact 1 in building up a picture of the spell victim.

FACT SHEET

GC-MS is a technique that combines Gas Chromatography and Mass Spectrometry to identify and measure the concentration of chemicals in a sample.

Pins

Seven brass pins were carefully bent to 'kill' them. Whether this knowledge was well-known by the person who made the bottle, or they heard by local hearsay that bent pins were always used, we shall never know.

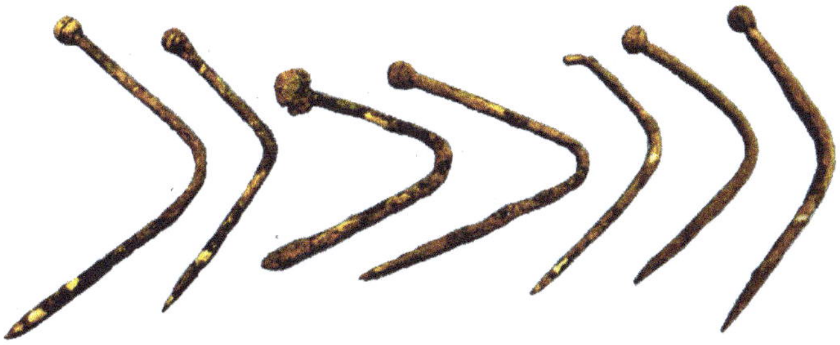

An SEM image of one of the pin heads found in the witch bottle.

Iron Nails

Historically, iron was thought to repel ghosts, evil spirits, fairies and witches, so it makes sense to include it in witch bottles. Twelve nails emerged from the Greenwich bottle, handmade from wrought iron and typically taper down their length. All but the two longest, stiffest ones were bent symbolically 'killed' like the pins and tufts of hair were entangled around them. By measuring the angles of the bends, Dr Massey was able to establish that the nails had been systematically pulled by pincers or a claw, from an already used piece of wood rather than being bent deliberately to be placed in the bottle. With the closeness of the River Thames to the burial site of the bottle, perhaps its maker had salvaged the nails from an old boat.

Eleven nails are pictured her, the twelfth has a more sinister feeling.

A Pierced Leather Heart

A slight shadow on the X-ray images was found to be a small piece of cow leather, measuring about 2 cm, cut into what could be described as heart-shaped and pierced with one of the hooked nails. Close examination of the leather 'heart' revealed that a slit had been cut first and the nail then pushed through. The average thickness of the leather was about 3 mm and may have been cut from one of the victim's shoes, these being intimate and personal items and one of the few articles of clothing that would have kept their shape when not being worn. Cloth hearts pierced with brass pins have been found in other witch bottles, but this is the first time a leather one pierced with a nail has been discovered.

Approximate actual size

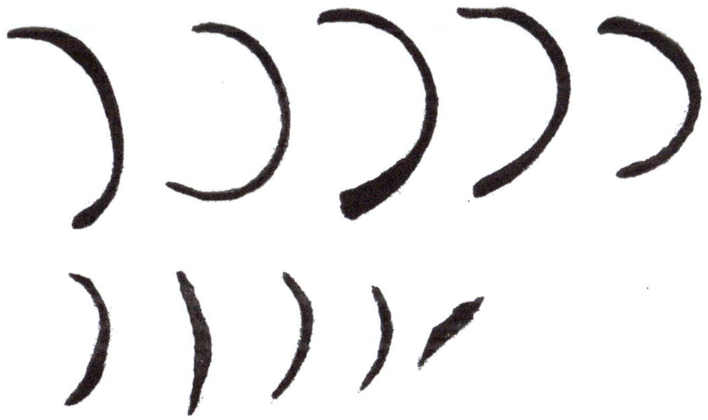

Fingernails

A quantity of brown debris, weighing 7.7 grams came out of the bottle and amongst the debris were found ten fingernail clippings. They were small in size with relatively undamaged outer edges and had been carefully cut using either a knife or a pair of scissors. These factors suggest that the victim may have been female and had not undertaken any heavy manual labour for some time. Was this because she had been ill for a long time, or she was from a higher social standing?

Fact 2 in building up a picture of the spell victim.

Microscopic examination of the debris, also revealed three cotton bundles, a rodent hair, possibly from a mouse, several feather fragments, cork debris and leather fibres from the 'heart'. All of these were coated in clusters of vivianite crystals, which is a hydrated iron phosphate mineral.

The SEM image shows how they have formed needle-like clusters.

Hair

There were three main bundles of hair in the bottle, each about 13-15 cm long. You can see the pins trapped in this dirty matted clump of hair, but after washing it several times with distilled water it was soon in much better condition.

Some shorter, loose hair fibres were also found in the debris from the bottle with many having frayed edges and two hairs having very acute bends. These hairs had most probably been pulled through the teeth of a comb.

So, what else was amongst the hair?

Urgh! There's Hair Lice in the Bottle

The broken and damaged bodies of eighteen hair lice and eighteen lice legs of various sizes were seen in the debris from the bottle. They were probably trapped in the teeth of the comb along with the broken hairs, and the act of combing would have resulted in their observed mutilations. Two of the louse's bodies were bent double, one forward and one backward, another had a broken back, and these injuries would have been caused by the comb teeth as it was pulled through the victim's hair. Their dark colour is significant, as lice tend to match the colour of their host's hair, so we can assume that the colour of the Greenwich hair is little changed from when it was deposited.

You could argue that head lice would make ideal additives to torment the witch, but to catch and add live lice would be very difficult, so it's safe to assume that the broken lice were attached to the comb from which the loose hairs were pulled. Modern combs are now made from a material that doesn't allow debris to collect between the teeth, but hundreds of years ago, combs quickly developed disgusting wads of broken hair and gunk which had to be cleaned off frequently. Old combs discovered in the tombs of ancient Egyptians have been found with lice in them, showing that these minute creatures have no respect for a person's wealth or status.

Infestations of hair lice were common, so the spell victim to whom this hair belonged should not be condemned too harshly. It's not easy to dislodge lice from hair, and simply washing it is not going to work. Lice can survive in water for several hours, and their legs have a pincer-like formation at the end of them, as you can see in the image below. They close the opposing thumb-like appendage on the claws, resulting in very secure anchorage to the hair.

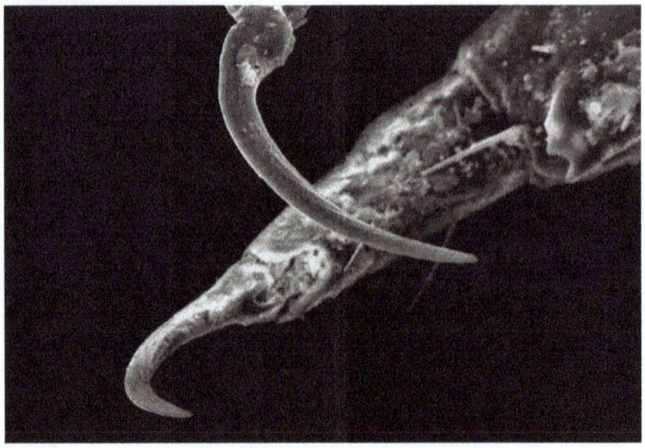

"There can be few people to be as pleased as I was to find their first hair louse!"

Alan Massey

Some of the hair fibres when they were examined under the microscope, had tiny segments of a white deposit on them, these are called peripilar casts and the longest one was 1 mm in length. These casts are caused when a hair does not completely separate from the hair root before it emerges from the scalp. Normally, they are destroyed by enzymes, but stress and some scalp disorders can disturb this process. They can also be caused by wearing tight hairstyles.

So, did the illness of the woman making up the bottle cause her to have a scalp disorder, or did she wear a very tight hairstyle?

Fact 3 in building up a picture of the spell victim.

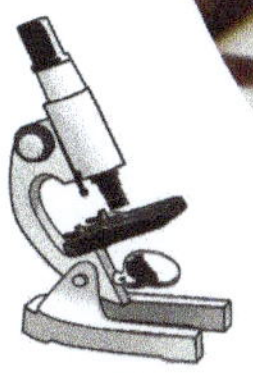

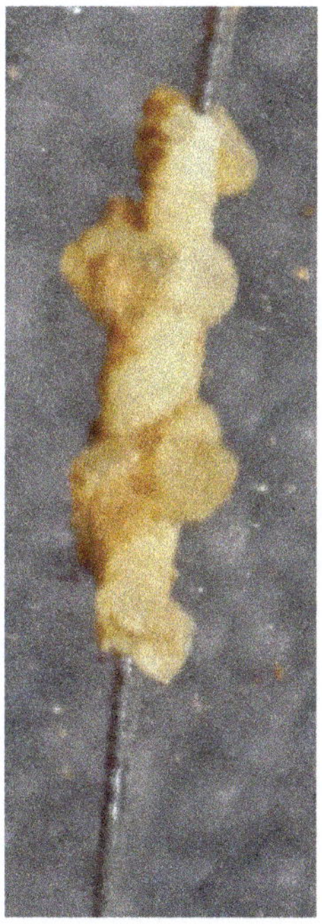

Peripilar cast under a light microscope

SEM of peripilar cast

"Small fragments of casts look very similar to nits (i.e. lice eggs) and hence they are often referred to as pseudo nits. Some of the Greenwich casts actually have lice eggs attached to them – a bizarre situation in which real nits are stuck to false nits."

Alan Massey

So, what did the science tell us about the spell victim?

It was a woman with dark hair who smoked and had head lice. She may have had a scalp disorder or worn her hair in a tight style putting stress on the hair follicles. The fingernail clippings suggested she had not endured hard manual labour or she may have been suffering from a long illness and not worn her nails down while she was ill.

Did the witch bottle work? That we will never know...

The Dorset Bottle

What strange goings-on were occurring in the villages of Langton and Worth Matravers in Dorset over 250 years ago?

During renovations of the old parish boundary, a corked glass bottle was found upside down within the footings, about 15 cm below the present ground level. It contained something that mysteriously seemed to change from solid to liquid when the bottle was handled, and this fact attracted particular attention. This phenomenon is known as thixotropic. The bottle, which may date from the mid-18th century, was imported and is made from pale green glass. Dr Massey was called upon to examine the bottle, and it arrived at his laboratory in July 1999.

What was Inside and What the Science Tells us

The contents were quite fluid and a dark brown colour, which made any visual examination of them impossible. A strong magnet was slowly drawn up the outside of the bottle, but nothing metallic came to the surface. When the bottle was opened, the smell issuing from it was quite nauseating! Inside was a dark brown mass that had the consistency of a very stiff syrup and stuck extremely tenaciously to anything it came into contact with. When ignited, it burnt with a bright, smoky flame typical of organic materials.

During the initial examination, the sticky mass was accidentally pierced with a spatula and revealed a lower liquid phase. The sticky organic material had permanently covered the whole of the inner glass surface so that the liquid layer could not be seen. It was only when the bottle was shaken during its journey to Loughborough that the liquid was able to break through the thick organic layer. The contents were not thixotropic as originally thought but two different layers, one organic and the other aqueous. It had only appeared that way when viewed through the almost opaque glass.

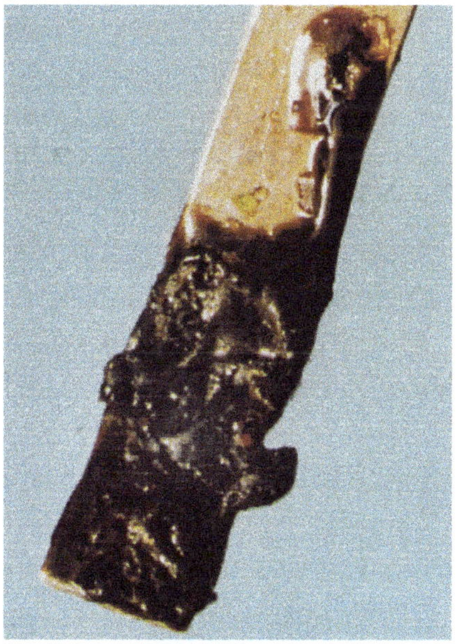

The Sticky Stuff

The Organic Layer

The sticky material was examined using a mass spectrometer, a technique which gives the molecular weights of the various elements in a mixture, thus allowing identification. So, what was it?

It was animal fat!

But what was in the aqueous layer? When a sample was left in the open air, it began to produce cube-shaped colourless crystals, which after analysis by infra-red spectrometry proved to be sodium chloride (salt).

The Aqueous Layer

A full nuclear magnetic resonance study showed the presence of glycerol, which had arisen from the animal fat. Further tests using chromatography and mass spectrometry intriguingly revealed nicotine. The final result of these tests concluded that the aqueous layer was cow urine!

So, why were animal fat and cow urine put in a bottle?

Perhaps the villagers wished to protect their cattle from disease from a neighbouring parish and buried it as a blanket protection for all cows in the district. But protection from what? In both 1714 and 1745, there was an outbreak of 'distemper of horned cattle', possibly rinderpest. The Dorset bottle is a rare example where a witch bottle can be linked to a historic event.

FACT SHEET

Nuclear magnetic resonance (NMR) spectroscopy is a non-destructive analytical technique that studies the physical, chemical, and biological properties of matter

A Bottle In A Grave

Perhaps the most bizarre witch bottle found is this small blue glass bottle, found by archaeologists in Loughton Churchyard near Milton Keynes in 1994. It was discovered inside a grave. Part of the graveyard was being cleared to make way for an extension to the church, an operation that required the careful relocation of some of the graves. When grave number 106 was opened, the skeleton was found to have a small bottle of liquid held between its left humerus and chest. The bottle is of a type once used by apothecaries to hold medications and, because of their shape, are known as 'steeple' bottles. This one dates to the early 18th century. The bottle is now stored at the Amersham Museum, Buckinghamshire, in a small, locked cabinet.

Actual Size

The bottle is 14.6 cm high, and its inside neck diameter is barely 6 mm, and although it may be a little bottle, it raises some big questions, not the least of which is, how many similar bottles are there in other graves? Who buried the bottle with the body and why? Were Church authorities aware of such an object being buried in consecrated ground? After all, witchcraft was a pagan belief. Was the bottle made up before or after the victim's death?

The bones were in too poor a condition to be able to determine whether they were male or female, but the age of death was estimated to be late teens to early twenties. One reasonable explanation for the bottle being in the grave could well be that the family of '106' wished the supposed witch who caused the fatal illness to suffer continued torment beyond the death of her victim.

Or…

> ***"Was someone with a deep-seated grudge trying to harm the spirit of '106' in the afterlife?"***
>
> Alan Massey

What was Inside and What the Science Tells us

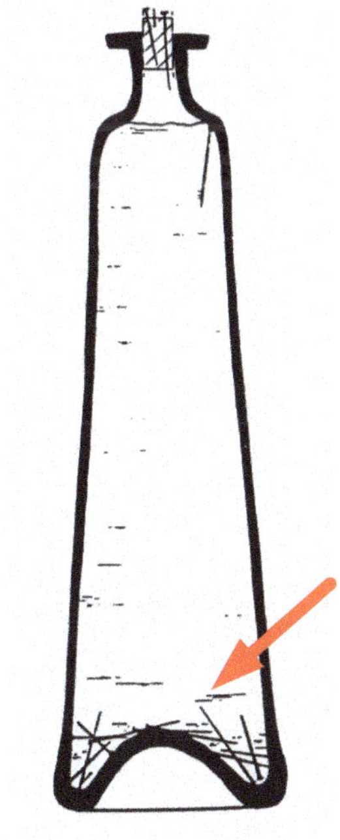

Once again, Dr Massey was able to examine the bottle, but unfortunately, he was not permitted to take samples of the liquid inside for scientific analysis due to the fragile condition of the bottle. However, when the door of the cabinet where the bottle is stored was opened, a strong smell of ammonia was apparent, suggesting that the liquid was urine.

The surface was quite corroded after centuries in the soil, making it impossible to see what was inside unless a powerful light was shone behind it. A copper-blue liquid was seen, but any solid objects were obscured by the large push-up at the base of the bottle. The archaeologist's report from 1994 stated that the liquid almost filled the bottle and that pins could be seen inside.

The Pins

So, what damaged the cork?

The small cork had been pierced with ten brass pins, one of which was headless and bent at almost a 90° angle. Presumably, this pin had been 'beheaded' deliberately before being pushed into the cork and bent double. A groove in the side of the cork shows where the tenth pin would have been but is now lost. It's incredible how the pins could have been made to pierce the tiny cork. The act of pushing these pins through the cork would have caused massive damage and must somehow thought to be connected with the idea of transporting the bottle to the spirit world where it could act out its magic.

The Liquid

What made the liquid turn blue?

Since 1994, half of the liquid has evaporated and been replaced with air due to the small cork in the bottle being damaged. The air slowly attacked the pins, resulting in copper corrosion products which have given the liquid its characteristic blue colour.

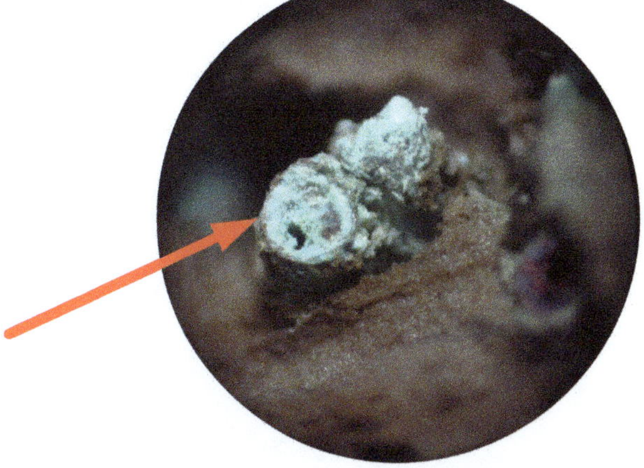

A Yorkshire Witch Bottle

The 'Old Farmhouse' in Newell-with-Clifton in North Yorkshire was first built in 1647, with an extension added ten years later. When a new owner began redecorating the house, to his surprise, he found no less than three blocked-up fireplaces hidden behind plaster and wallpaper. When he broke through the brickwork of one of them, he was even more surprised when a small brown, earthenware bottle fell out of the chimney. From dating one of the bricks, it can be estimated that the fireplace was probably blocked up during the late 19th to early 20th centuries.

Dr Massey's expertise was called upon once more, and the scientific analysis began.

The bottle measures about 16.5 cm tall and is half-covered with a thick, dark brown glaze. The X-ray image shows the nails and pins secreted inside.

What was Inside and What the Science Tells us

But First, What was on the Outside of the Bottle?

The Dust

On examination of the bottle, it was seen to be coated with a grey dust. A sample of this dust was analysed by powder X-ray diffraction (XRD), which is a rapid analytical technique that uses X-rays to study the crystalline structure of materials. The grey dust was mainly calcium sulphate with a small amount of quartz. Presumably, sulphur dioxide produced during the burning of coal had, in the presence of moisture and air, attacked lime mortar in the chimney to form the calcium sulphate. Over the years, it had drifted down the chimney and coated the bottle. The quartz (sand) was released from the attacked mortar and contributed to the coating. The dust had settled evenly over the bottle lip, proving that the bottle had been placed upright in the chimney, possibly on a shelf or ledge. The cork which initially sealed the bottle had long dried up and shrunk from the hot flue gases and fallen inside.

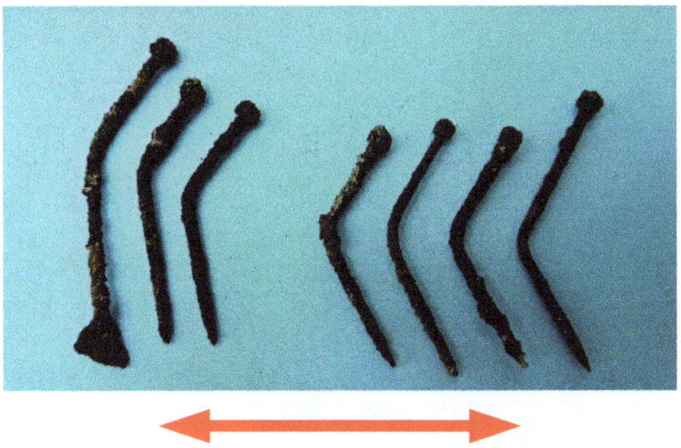

4 cm

The X-ray image of the bottle had clearly shown up pins and nails, and when the contents were shaken out into a petri dish, seven corroded pins and eight very rusty nails were picked out.

The Pins

The pins were all bent, and the green patina on them revealed they were made of brass. The rounded shape of the pin heads showed they were handmade, as machine-made pins were not common until about the 1840s. The deliberate bending of these pins once again reinforces the suggestion that this was a witch bottle.

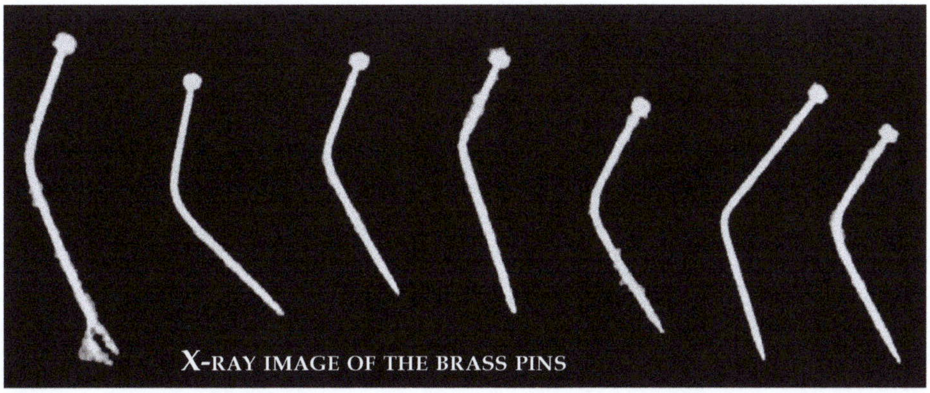

X-RAY IMAGE OF THE BRASS PINS

The Nails

The short nails are completely flat with no pronounced point, being of a type called 'cut nails'. Their shape suggests they could have been made around 1800, which gives an approximate date for the filling and deposition of the bottle.

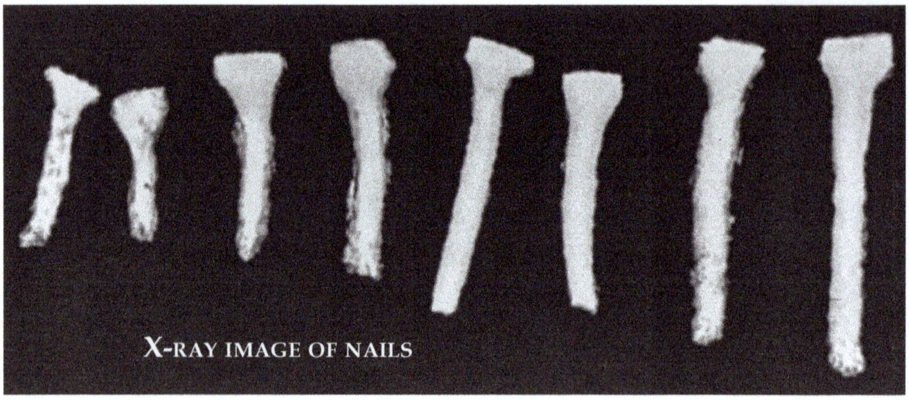

X-ray image of nails

Urine

When the debris from the bottle was shaken through a 1.5 mm sieve, 12.9 grams of soot passed through, and to test to see if any urine had once been present, a sample of the soot was extracted with warm water, filtered and set aside to evaporate. The crystals that formed were analysed by infrared and XRD. The results identified calcium sulphate, ammonium chloride, sodium chloride, and hydrated calcium phosphate, and, along with colourless lumps of calcium carbonate picked out of the soot, they are all compounds expected to be found in decayed urine.

Dust Bunnies

Under the microscope could be seen a complex mixture of charcoal fragments, hair, dust particles, white feathers, tiny grass stalks and a fly's head. These had gathered together into balls known as 'dust bunnies'. Several hair fragments were in the debris, some free and others trapped in the dust bunnies. The hairs are assumed to be human, and some have their ends square-cut, perhaps suggesting they were cut with scissors. The big question, though, is how did these hair fibres get into the bottle? The shorter ones might have simply floated in over the years, but the long hairs would have great trouble doing so, especially against any upward draught in the chimney. On the other hand, there seem to be far too few fibres for a lock of hair to have originally been placed in the bottle.

The Beetles

When the debris from the bottle was examined under a low-powered microscope, dozens of tiny dead beetles were seen. They were identified as ptinid beetles, each about 3 mm long.

What were they doing inside the bottle?

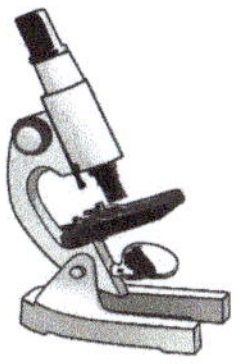

Initially, they may have been attracted to old bird's nests or similar material higher up the chimney, and when they fell into the bottle, they would have died very quickly from the foul atmosphere. Rather charmingly, these little beetles allow us to recognise three distinct episodes of the bottle's history when it was inside the chimney.

Phase One

The corked bottle containing urine, brass pins and iron nails is placed up the chimney, all the available evidence suggests it was upright. Fires can be lit because the hearth and chimney are open.

Phase Two

The cork has now rotted, shrunk and fallen inside the bottle. Tiny ptinid beetles and a housefly enter the bottle where they are exposed to fumes from the fires and become blackened, losing their heads, legs, antennae and hair in the process. All the urine is lost by evaporation early on, leaving the brass pins and iron nails open to the corrosive effect of damp air and acidic fumes. The blackened ptinid beetles prove that the bottle had been in the chimney before the fireplace was blocked up.

Phase Three

The fireplace has been bricked up, but the open bottle is still collecting airborne debris. All of the ptinid beetles which enter now retain their golden colour, many legs, antennae and hair; a few other types of small beetles and a housefly are also captured. Thus, the excellent condition of these new ptinid beetles proves uniquely useful in defining the division between phases two and three.

"No fingernail clippings or other personal items were found during extensive microscope searches of the solid debris, so it has to be assumed the bottle was not intended to cure someone of a witch's spell. The placement in the chimney, from where 'evil spirits' might enter, points to protection of the house being intended."

Alan Massey

A Sad Little Bottle From Suffolk

How little evidence is required to claim that an object was actually a witch bottle? This question had to be asked of a little glass vial measuring 7 cm tall with a 2.1 cm diameter, found hidden near the gable chimney of a house in Great Back Lane, Debenham, Suffolk. It probably dated from the 18th century and was a somewhat misshapen bottle that was probably free-blown rather than moulded. The fragile lip is intact, but there is a chip missing from the rather thick base, which may have resulted from a deliberate attempt at a symbolic killing.

Actual Size

This tiny bottle must have been laid horizontally in its hiding place near the chimney because a deposit of greyish solid material about 12 mm wide extended from the neck to the base and, crucially, covered the chip that was struck off the base. Clearly, the chip occurred at some time prior to the bottle being hidden. Analysis of the deposit showed the presence of chloride, nitrate, sulphate, phosphate and probably acetate, all of which would be expected components found in chimney dust.

"There is such a sad air of poverty about this bottle, as well as it being a sign of someone's illness."

Alan Massey

The Bottle Lip

What was Inside and What the Science Tells us

The original cork had long since rotted away and was replaced with a new one in the 1970s. As this cork was gently removed, a hair, almost certainly human, was noticed stuck firmly to the inside of the neck, and a few more hairs were inside the bottle. The neck aperture was so tiny, about 12 mm in diameter, that it is improbable these hairs could have floated in accidentally, indeed, it had obviously been difficult even to push them inside manually, as evidenced by the hair stuck in the neck.

Hair

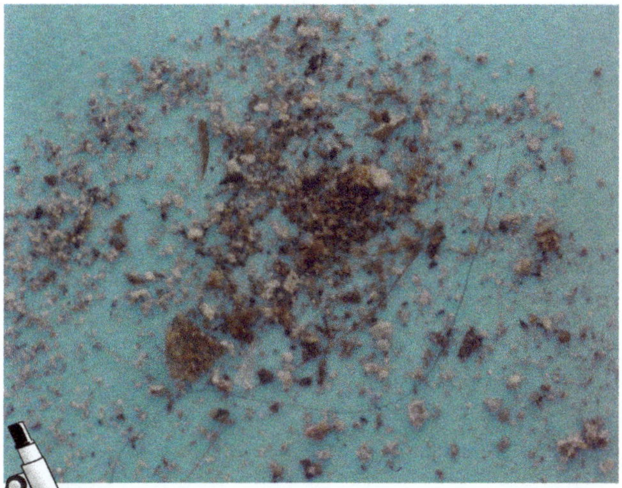

Debris

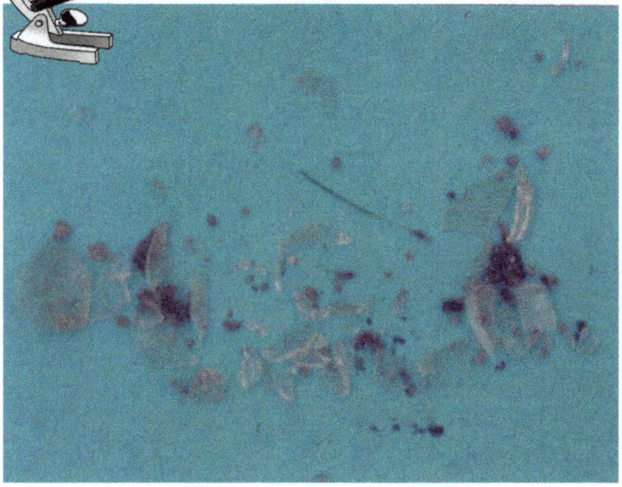

Glass Fragments

The Debris

The meagre contents which were shaken out of the bottle contained bits of rotted cork, a few silica grains typical of soil, and glass shards, probably from the damaged base rather than being deliberate additives.

Urine

When the water used to flush out the bottle was analysed, it contained chloride, sulphate, nitrate and phosphate anions, all of which are typically present in decayed urine.

What story did this bottle tell?

Could this be the smallest witch bottle on record? It seems probable it is, but perhaps one made up by an extremely poor person who could not afford a better receptacle or the usual metal additives. Significantly, the tiny vessel, corked with hair and urine inside, had been carefully hidden near a chimney. Doors and chimneys were dangerous openings where evil spirits might enter a home and so must be guarded.

A Witch Bottle From Exton, Rutland

Let us imagine for a moment the following scenario, happening some three and a half centuries ago in an age when superstition was rife, veterinary skills and medical knowledge were modest, and antibiotics were yet to be discovered… He had lost three of his valuable horses in as many days. What was wrong? Who had cursed him so cruelly? As his new cottage took shape, he decided he must seek help from a local cunning man or wise woman in protecting the rest of his stable; the words whispered in his ear that day were known to no one, but their manifestation was discovered over three centuries later – hidden beneath the great hearth of his cottage. Whether the magic device had worked or not, we can never know…

The drawing shows the rear of the cottage where the bottle was found c. 1916.

Like many old properties, a 17th-century Rutland cottage was being renovated in 1983, and a bellarmine bottle was discovered, unusually, standing upright beneath a large fireplace. It was sealed and apparently contained liquid, judging from the motion felt when it was shaken. The owner of the cottage alerted the find to Rutland County Museum. The first scientific study began in 1984, but Dr Massey did not receive its contents at Loughborough University until eighteen years later for him to carry out further scientific analysis.

The bottle is a typical stoneware bellarmine, measuring 20.5 cm tall, and can be dated as being made from c. 1650-1675. The whereabouts of the bottle now are unknown, and luckily, the image above was taken just after its discovery in 1983.

What was Inside and What the Science Tells us

When the bottle was examined initially, the study started by X-raying the bottle to see if any metallic objects were inside, but none were present. Granular material sealed the neck, and after it was removed, the muddy brown liquid that was poured into a glass beaker smelt rather like a farmyard. Its pH was 7.4, and sodium, potassium, calcium and magnesium cations were present, but the big surprise was a high concentration of oxalate anions. The presence of these was not understood for some years. The grey solid from the bottle, when dried and examined under the microscope, revealed crystals of calcium oxalate.

Urine

When Dr Massey began his study, virtually all of the oxalate in the liquid had decomposed, however, calcium oxalate was still present in the grey solid as determined by analysis by powder XRD. A full analysis of the liquid was carried out, the results of which are consistent with a decayed urine sample which had originally contained considerable amounts of oxalate cations. Horse urine has these properties, and furthermore, it deposits crystals of calcium oxalate formed on standing.

Horse Fat

Samples of the grey solid were extracted in different organic solvents and analysed by infrared spectrometry and GC-MS. What we had in the witch bottle were types of fats known as triglycerides, and the two identified were stearic acid and palmitic acid, the ratios to each other of which are found in horse fat.

Why would someone make a witch bottle from horse fat and urine?

The address where the bottle was found was Maltings Yard, and a maltster would have required a team of horses to ply his trade. Historically, there does not appear to have been any widespread 'plague' affecting horses, however, one very contagious disease affecting horses, called 'glanders', had been around for centuries but now, thankfully, is eradicated in this country. Possibly, the bellarmine and its contents were considered as insurance against glanders attacking the maltster's stable.

"There is little doubt that this bellarmine was concocted with the curing or protection of horses in mind."

Alan Massey

FACT SHEET

A solvent is a substance that can dissolve an insoluble solute to create a solution. An organic solvent is an organic liquid that vaporises at room temperature and they each have a distinctive smell. Examples are acetone, ethanol and methanol and many organic solvents are used in everyday products like paint, varnishes, glues etc.

An 'Unfathomed' Bottle from Trispen, Cornwall

This particular bottle was identified in a 2016 episode of Antiques Roadshow as being a port bottle from the 1800s. It still contained liquid, and the presenter, Andy McConnell, proceeded to draw a sample of the 'very brown' liquid through the cork with a syringe and drink it. He remarked, "I think it's port – port or red wine... or it's full of rusty old nails and that's rust." In a later episode, in April 2017, it was revealed that the liquid was in fact urine, and thirty-five brass pins were discovered inside the bottle. It had been found upside down under the doorstep of a cottage on Trispen, Falmouth, in Cornwall, and the facts all pointed towards it being a witch bottle.

Dr Massey was able to examine the bottle contents further, and the investigation began in February 2018.

The X-ray shows the pins laying in the bottom

What was Inside and What the Science Tells us

The bottle was totally opaque and had been fitted with a modern cork after the original one had been accidentally pushed inside the bottle when attempts were made to extract a sample of the liquid to use on the Roadshow programme. A powerful torch placed behind the bottle showed it to be half-filled, but little else was visible apart from numerous air bubbles trapped in the glass during manufacture. The owner of the bottle had warned that the contents had a rather bad smell, so the cork was removed in a well-ventilated room and the liquid poured into a beaker. Altogether there were 200 mls of a brown, almost opaque liquid, thirty-five pins and a quantity of solid debris.

The Pins

Unusually for a witch bottle, the pins were straight. They were machine-made and so can be dated to about 1840, ranging in length from 21 to 40 mm with several having a prominent seam running lengthwise due to the style of manufacture. The shafts had been made by passing brass sheet through dies rather than solid rod. These seams formed when the two edges of the sheet met and tended to be areas where corrosion appeared. Many pins had a solid mass of deposited crystals attached to them.

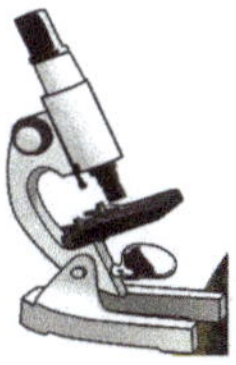

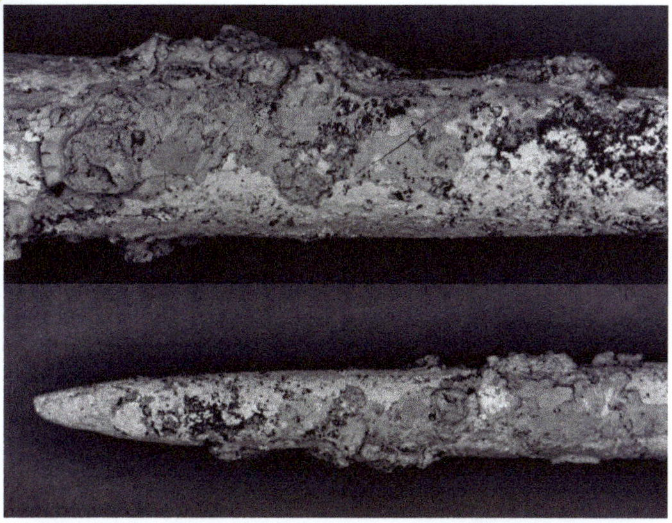

SEM images of the corroded pins

The Debris

Within the dried bottle debris are fragments of stalks and grape skins which must have been small enough to pass through the sieves and filters during clarification of the port. Addition of acetone to the skins gave a reddish-pink solution, amazingly, the grapes' pigment had been retained from the vine, through port production, and the years in the witch bottle, right up to the present day! Close examination of the solid debris revealed a single human hair fibre, which carried tiny clusters of crystals, coloured slightly by the port, so clearly it had been inside the bottle since deposition and was not a modern contaminant. The thickness of the hair points to a male owner, and almost certainly, he would be the person who had dug the hole under his doorstep to hide the port bottle.

FACT SHEET

1. A colloidal suspension is basically a liquid in which tiny solid particles float and do not dissolve.

2. An inductively coupled plasma emission spectrometer (ICP-OES) is a device that uses a plasma to analyse the composition of elements in a sample.

The Liquid

Was it urine?

The concentrations of sodium, potassium and magnesium in the liquid are about one hundred-fold lower than expected for neat urine, and no calcium carbonate, a major urine marker, could be identified in the debris. A small sample of the liquid was filtered to remove much of the suspended matter, however, the brownish filtrate still contained a colloidal suspension which refused to settle out on standing. The filtrate was analysed by nuclear magnetic resonance (NMR), but the various components present were at an extremely low level, but two chemicals could be positively identified: benzoate and ethyl alcohol.

Benzoate occurs in old urine, but unfortunately, it's also found in port, so it cannot be used as a definitive urine marker. Analysis of the filtrate using an inductively coupled plasma emission spectrometer showed that sulphur was, by far, the main chemical. The big question, obviously, is how did this sulphur get into the bottle? First guess is that a pinch of brimstone had been added to 'spice up the magic' as is usual in a witch bottle, but, remembering the presence of port dregs in the bottle, there is another more probable source. Sulphite preservatives are added to wines, therefore, the probable origin of the sulphur.

So, what is the liquid? The probable answer came in a most strange manner...

The Ostracod

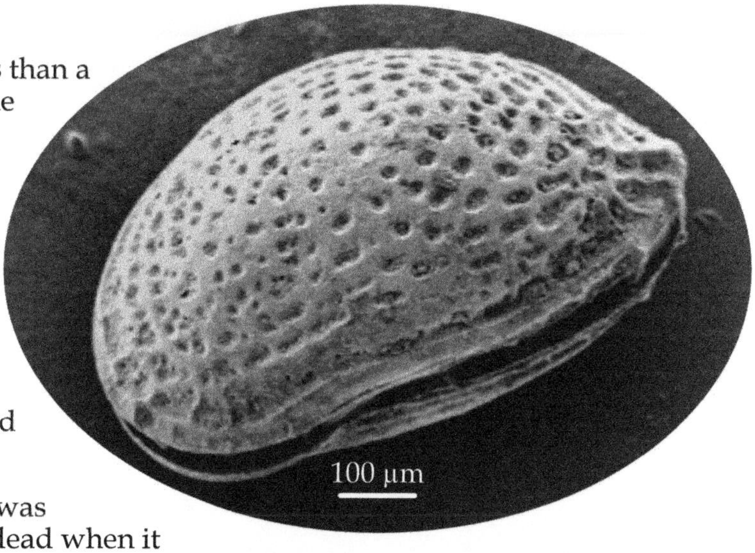

Under the microscope, one strange item, less than a millimetre long, stood out from the rest of the solid material. It looked like a tiny cockle whose shell was covered with dimples. It was identified as an ostracod, a tiny sea creature. They can be found living in British southern coastal waters, from the south coast of England, the Bristol Channel and the Welsh coast to Anglesey, down the French, Spanish and Portuguese coasts and into the Mediterranean. They are fairly common on seaweeds in rock pools and shallow coastal waters.

After careful examination of the ostracod, it was concluded that it was alive or only recently dead when it entered the port bottle. At only 0.73 mm long, it was far too small to have been captured and used as a deliberate bottle additive. The only reasonable explanation is that seawater was added to the bottle and the unfortunate ostracod was caught up in the sample. The SEM images show the incredible structure of this minuscule creature.

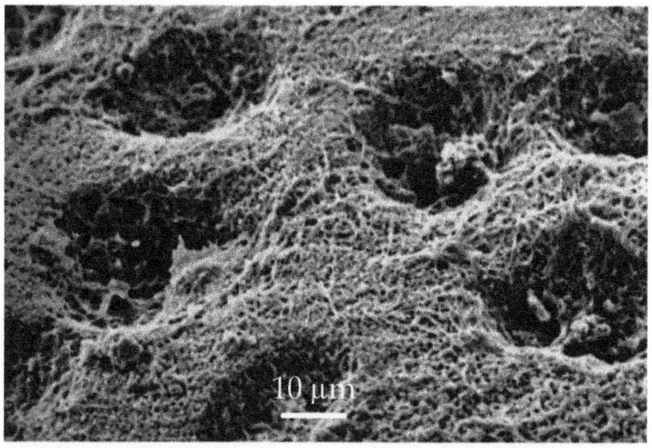

The Seawater

Why did the cottager go to such trouble in filling and then burying the bottle?

Urine and freshwater were readily available, and yet seawater was chosen to put in the bottle, despite the extra effort required to collect it. The preparation of the port bottle must have been a very important ritual, as Trispen is ten kilometres from the coast. This seems to imply that seawater was considered very important, perhaps to aim all the bottle's power at its desired target - the sea. Why?

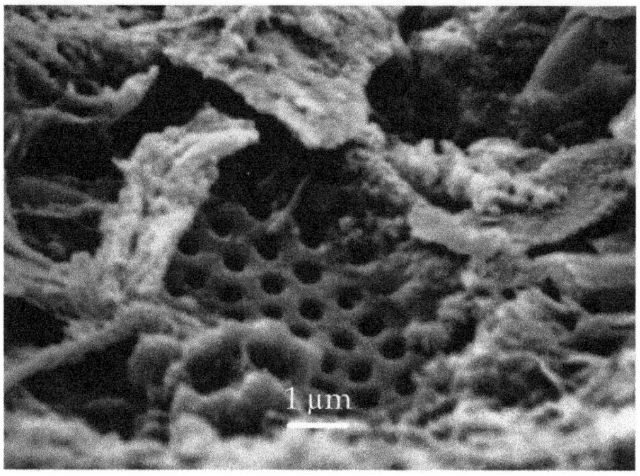

Perhaps the terrifying event which occurred in 1859 can provide a plausible answer. On 25th October, the whole of the western coast was visited by a violent storm. The wind whipped up the sea to such a fury that 133 ships were lost, many others damaged, and over 300 people lost their lives.

"We can never know for sure if the intended function of this bottle was indeed to give protection against future stormy seas. The evidence for this interpretation rests heavily on the accidental inclusion of a 'beast' not quite one millimetre in length – the ostracod. Such a storm would make a great and lasting impression on someone living in Cornwall. As the bottle was manufactured in the early 19th century, and the pins were of a type available in quantity from about the 1840s, the dates fit for the bottle to have been used as a defence against the 'spirit of the seas', giving protection to the house against further storm and tempest."

Alan Massey

Acknowledgements

Special thanks to Steve Wells for the front cover witch illustration and the 'blame it on the witch' illustration page 6.

Special thanks to Helen Nall for the front cover fireplace, nails and witch illustrations.

Crazy horse cartoon page 29 - Open Clipart by samtux

And of course, this book would not have been possible without Dr Alan Massey's impressive body of research and his kind permission to publish this abridged version of his work.

Alan and Sylvia Massey making a cup of tea.

A few months of retirement idleness, as Alan put it, had passed when he spotted an old newspaper article with a picture of archaeologist David Williams, holding the onion-shaped wine bottle he had dug up in Reigate. In the other hand, he had a goblet and pretended to drink the bottle contents. As Alan read further, he learned that the liquid from the bottle was urine that had been passed centuries earlier. His thought of, "What was ancient urine doing in a wine bottle?" aroused his interest and curiosity. Having retired from Loughborough University with a university fellowship and access to a whole laboratory of modern scientific equipment, here was an unprecedented opportunity to examine a 300-year-old sample of body fluid. And so began twenty-five years of uncorking the murky mysteries of witch bottles.

www.ingramcontent.com/pod-product-compliance
Lightning Source LLC
LaVergne TN
LVHW070523061126
851857LV00023B/2135
* 9 7 8 1 7 3 8 5 5 6 8 6 1 *